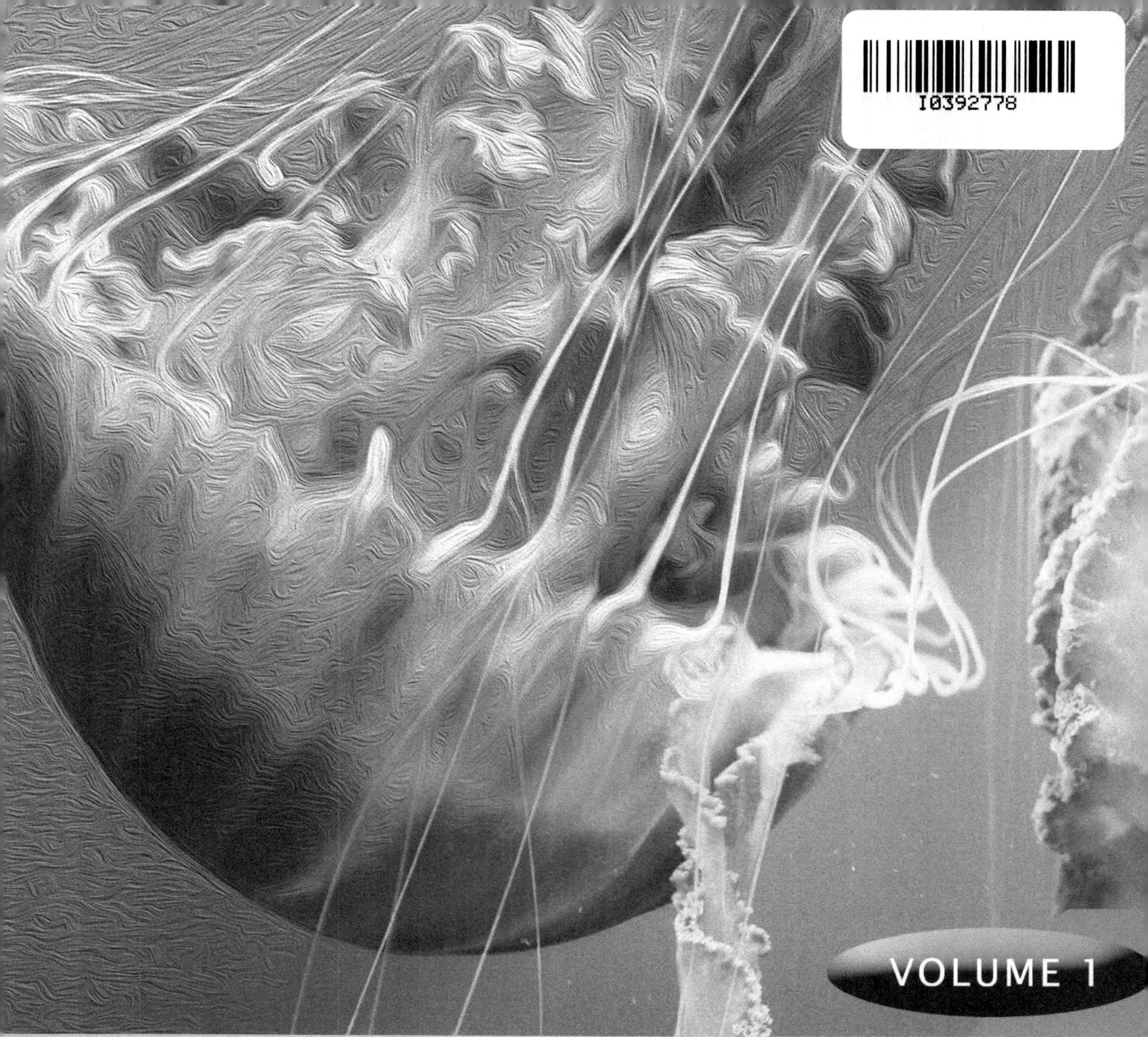

Sea Creatures
SHADING

A Grayscale Coloring Book

COLORING BOOK FOR ADULTS

VOLUME 1

COLOR TEST PAGE

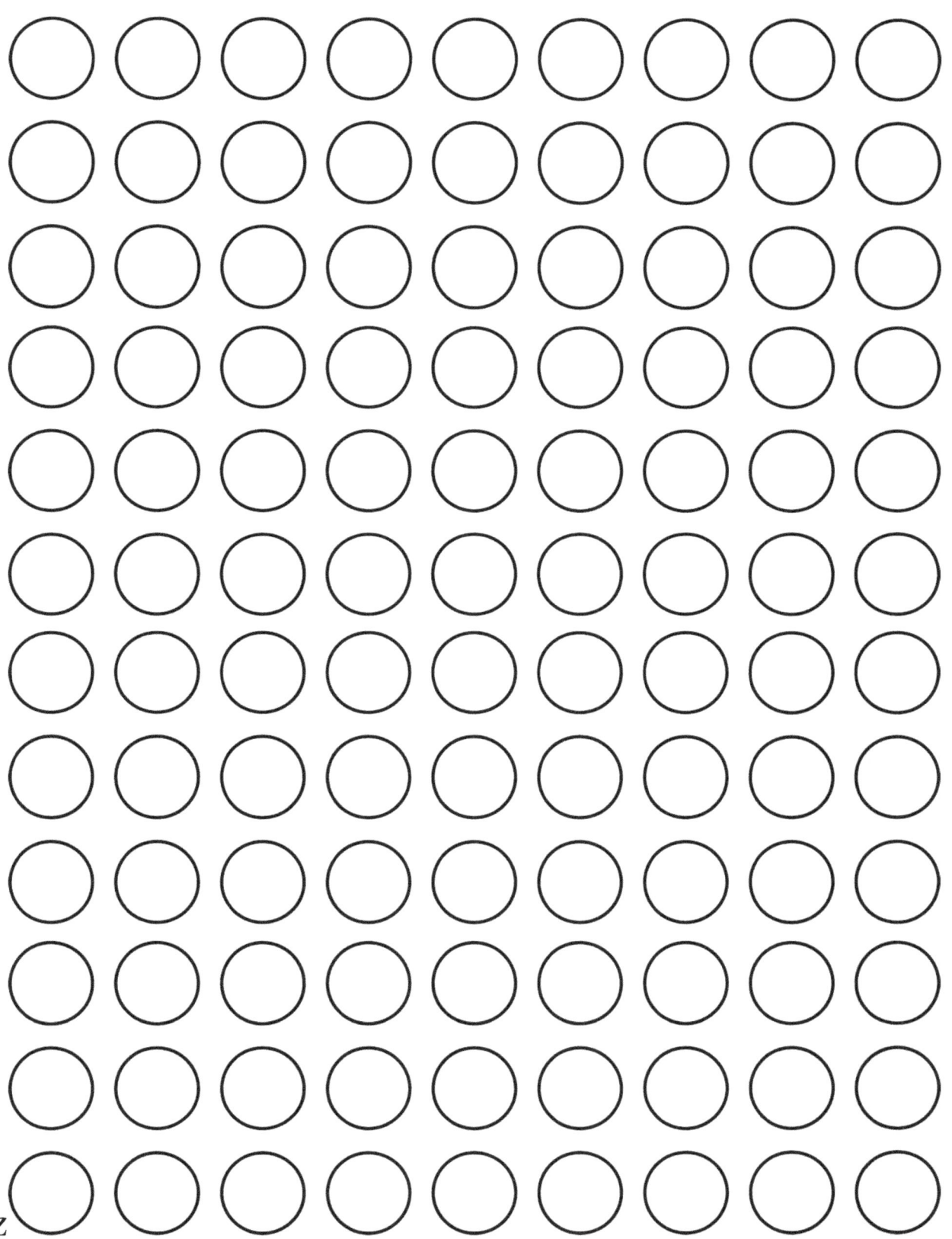

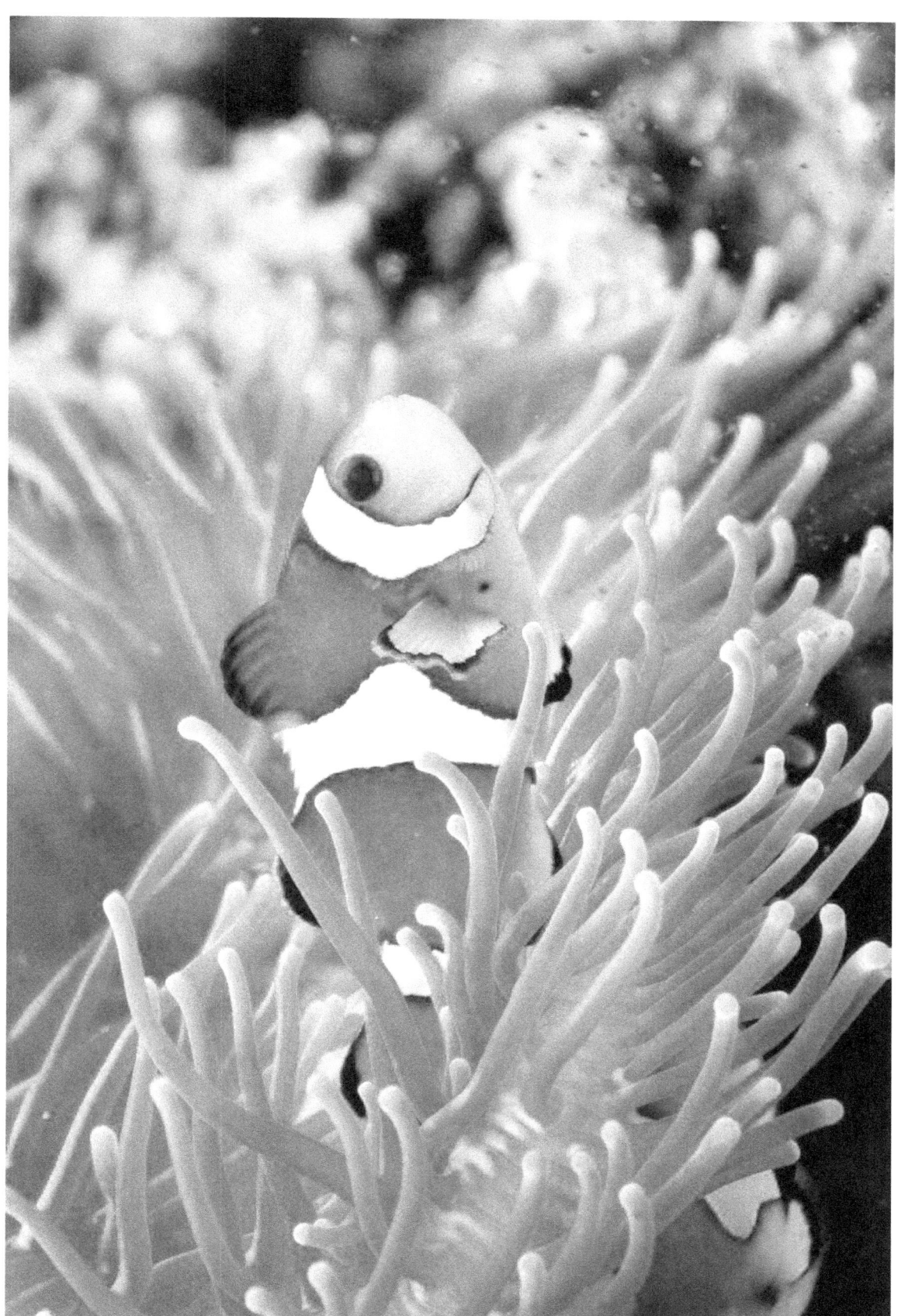

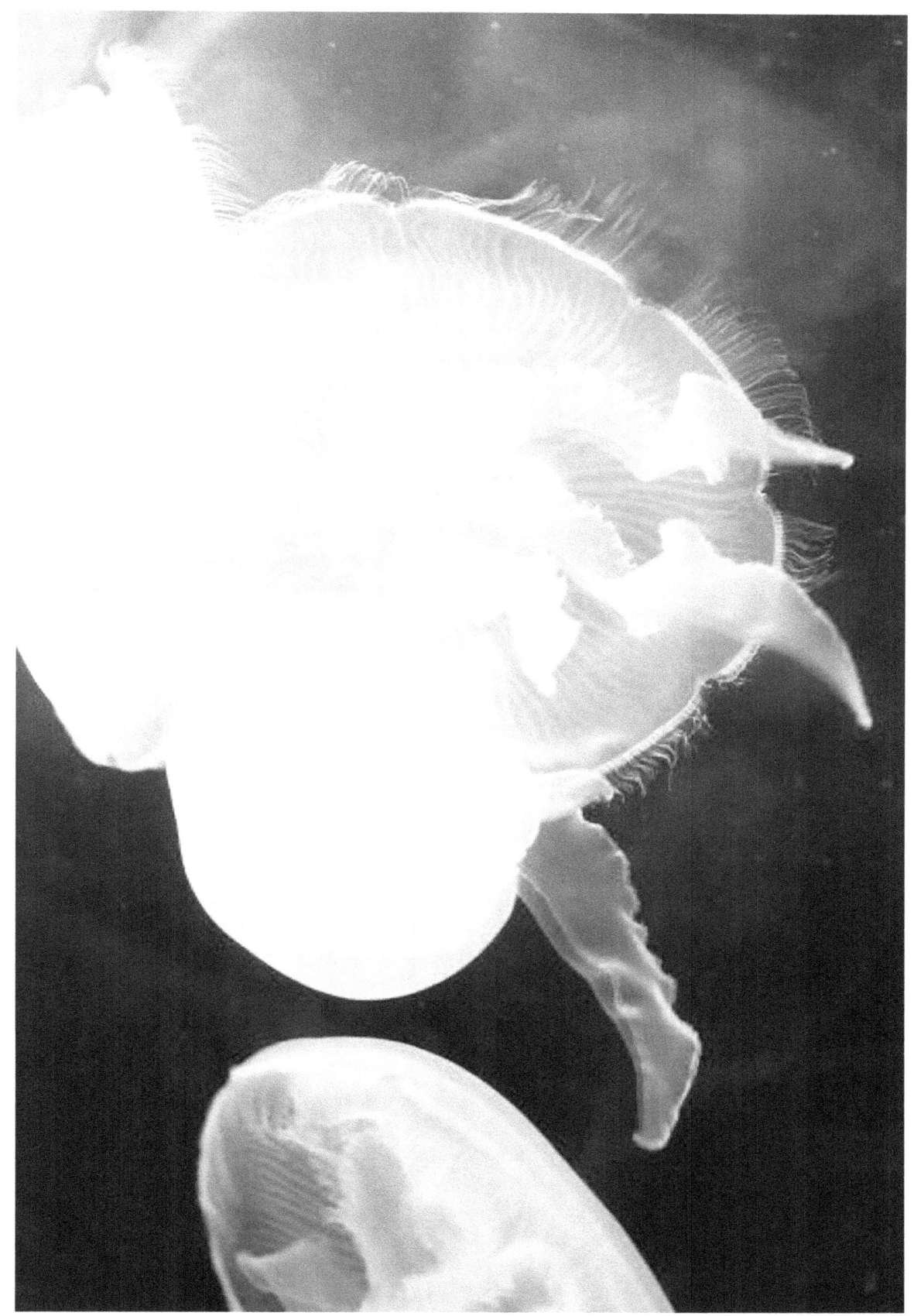

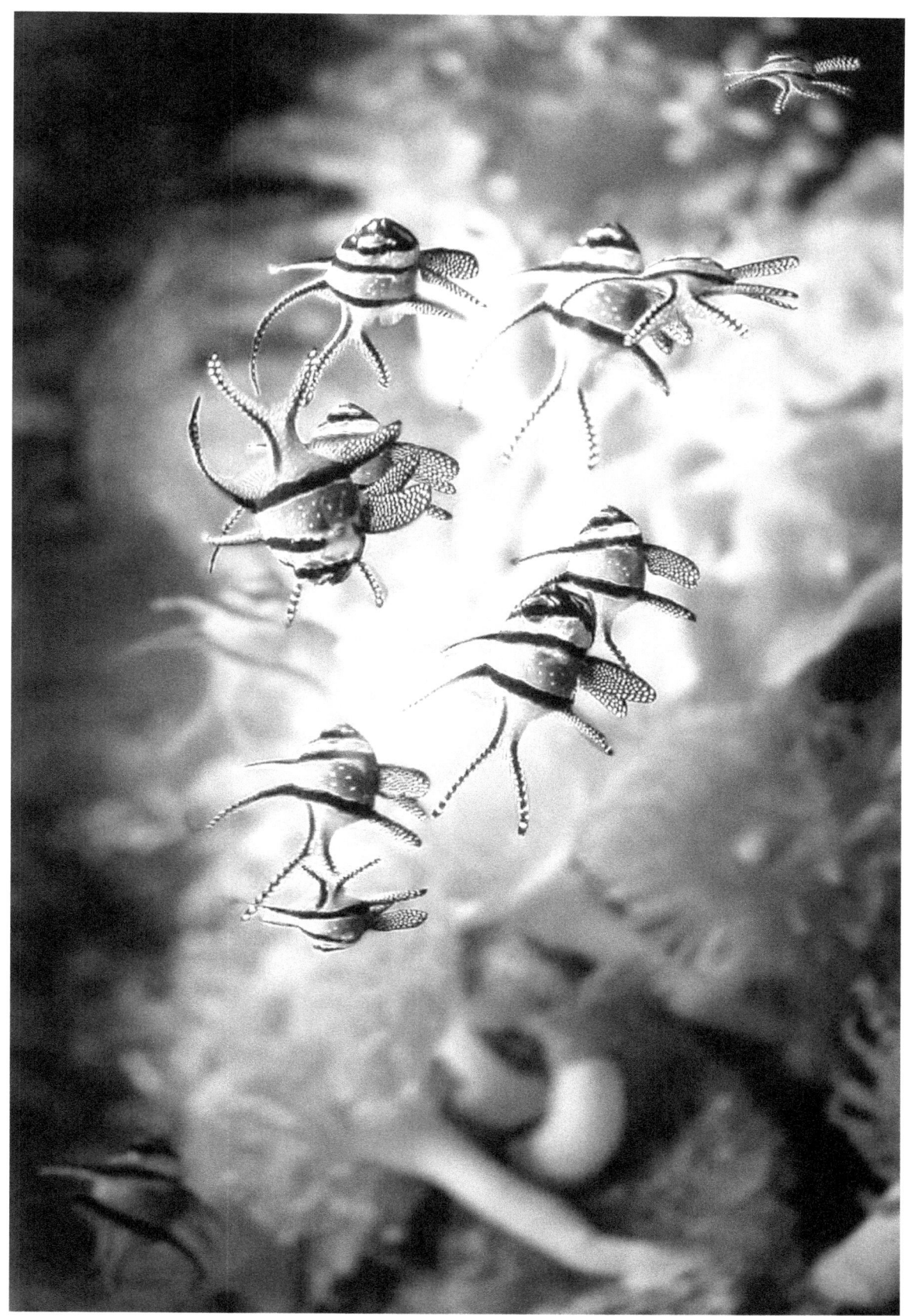

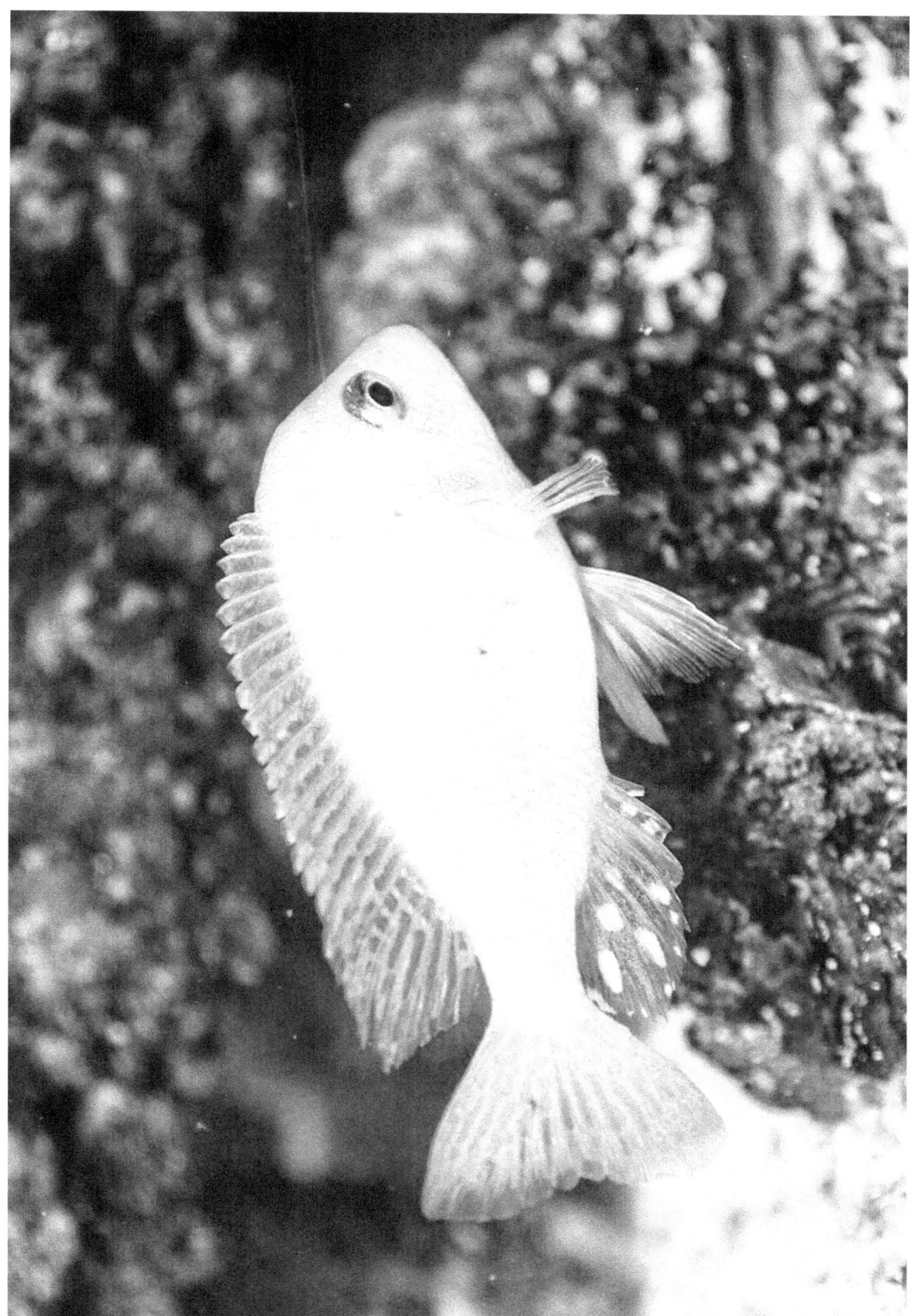

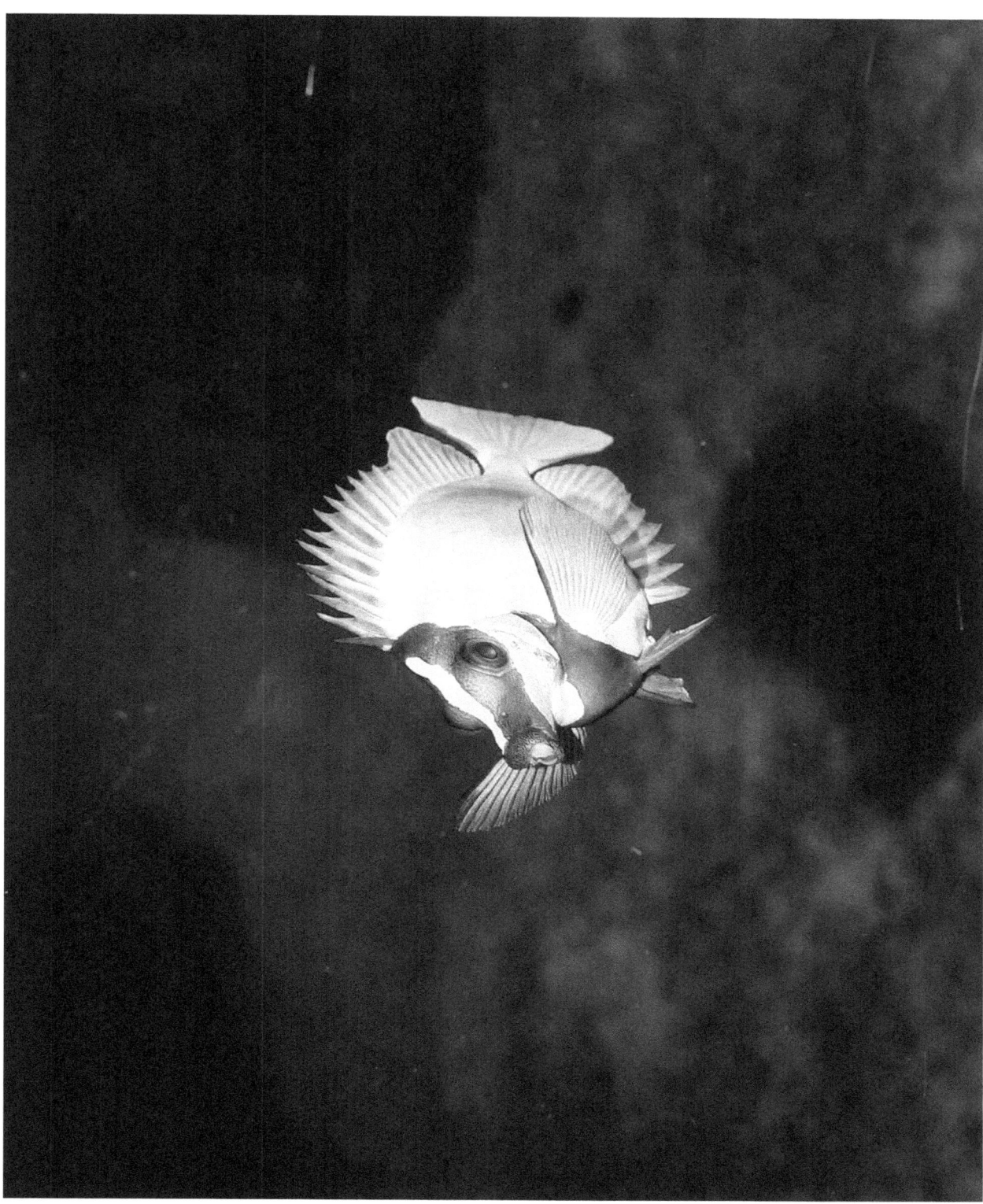

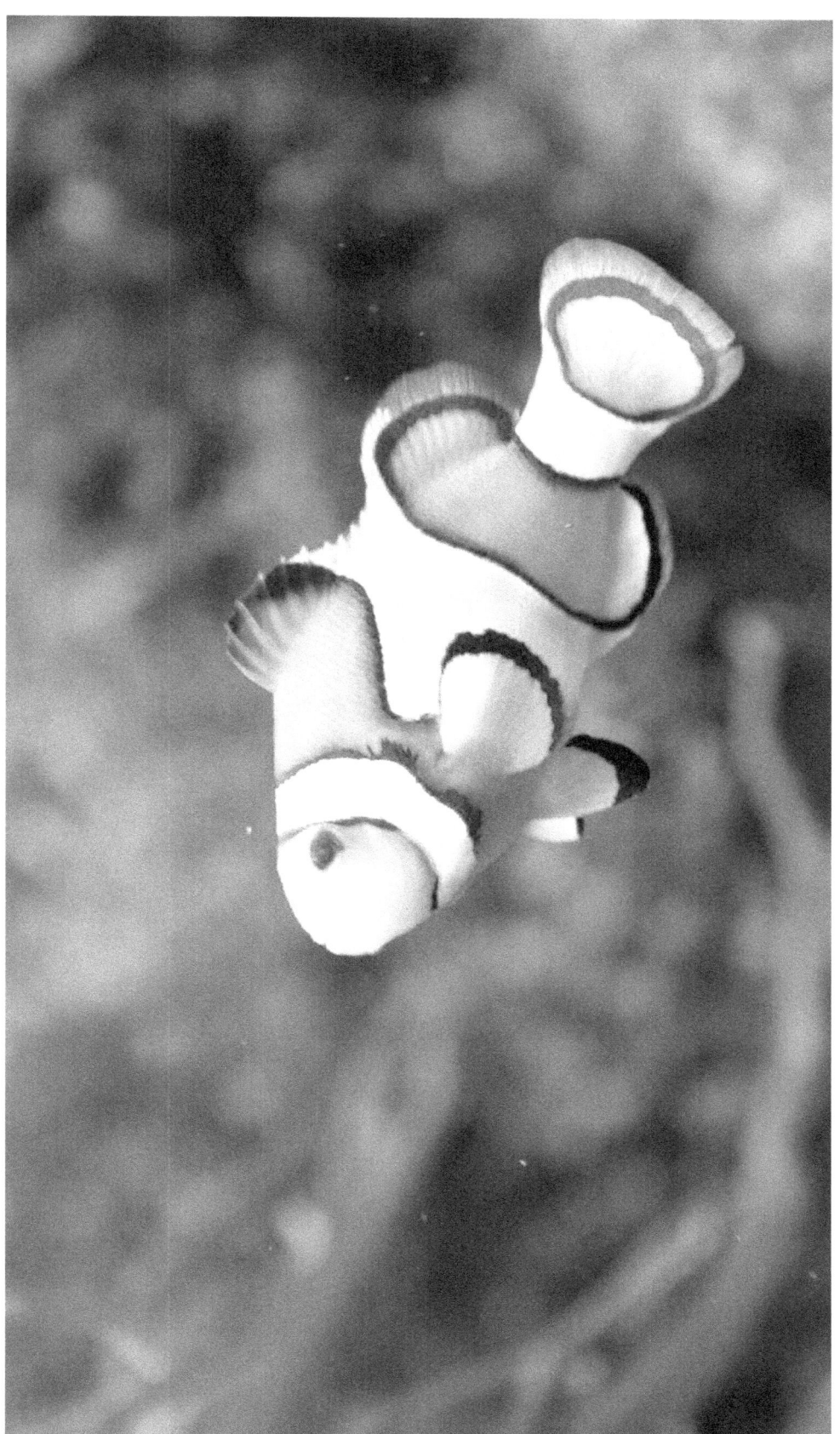

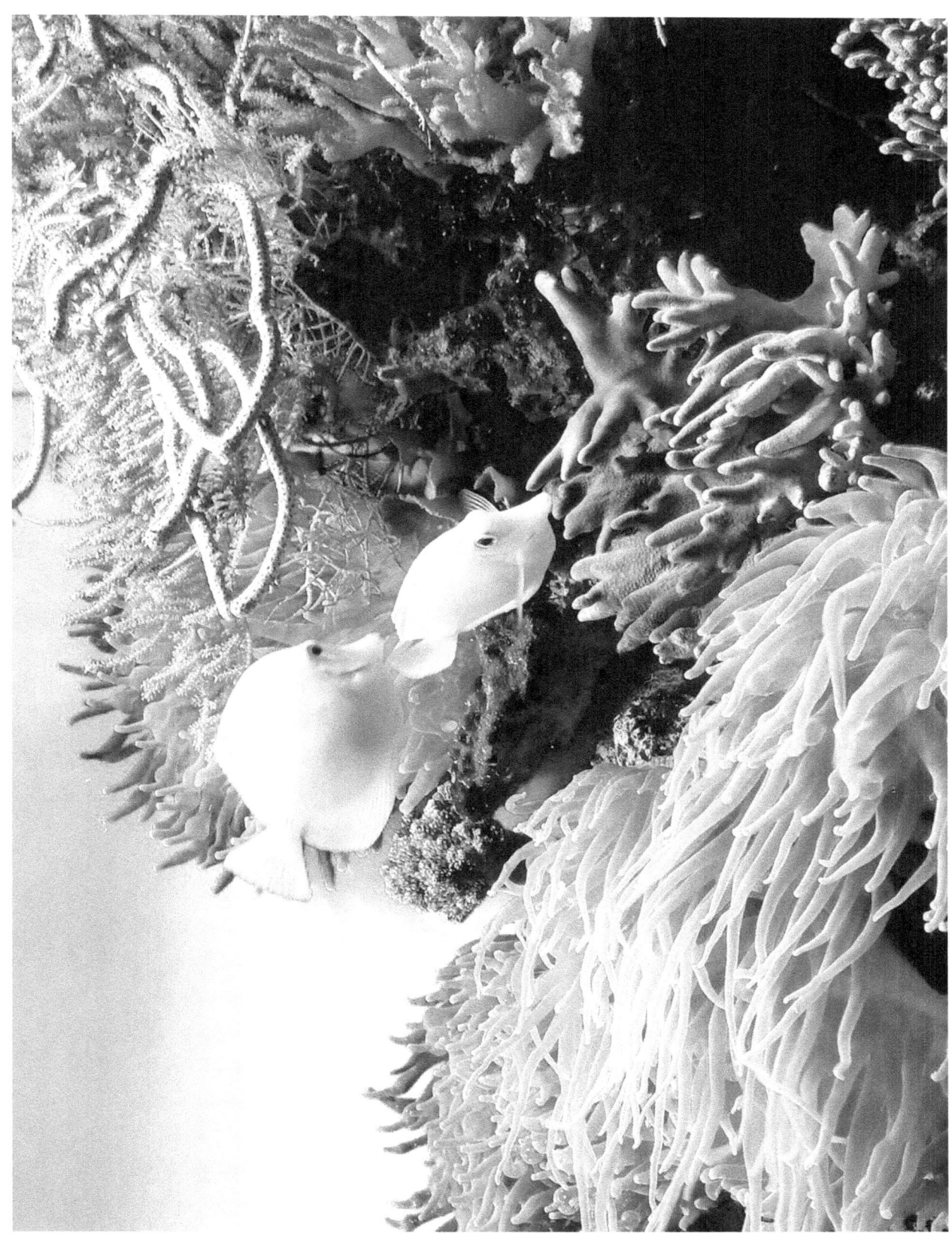

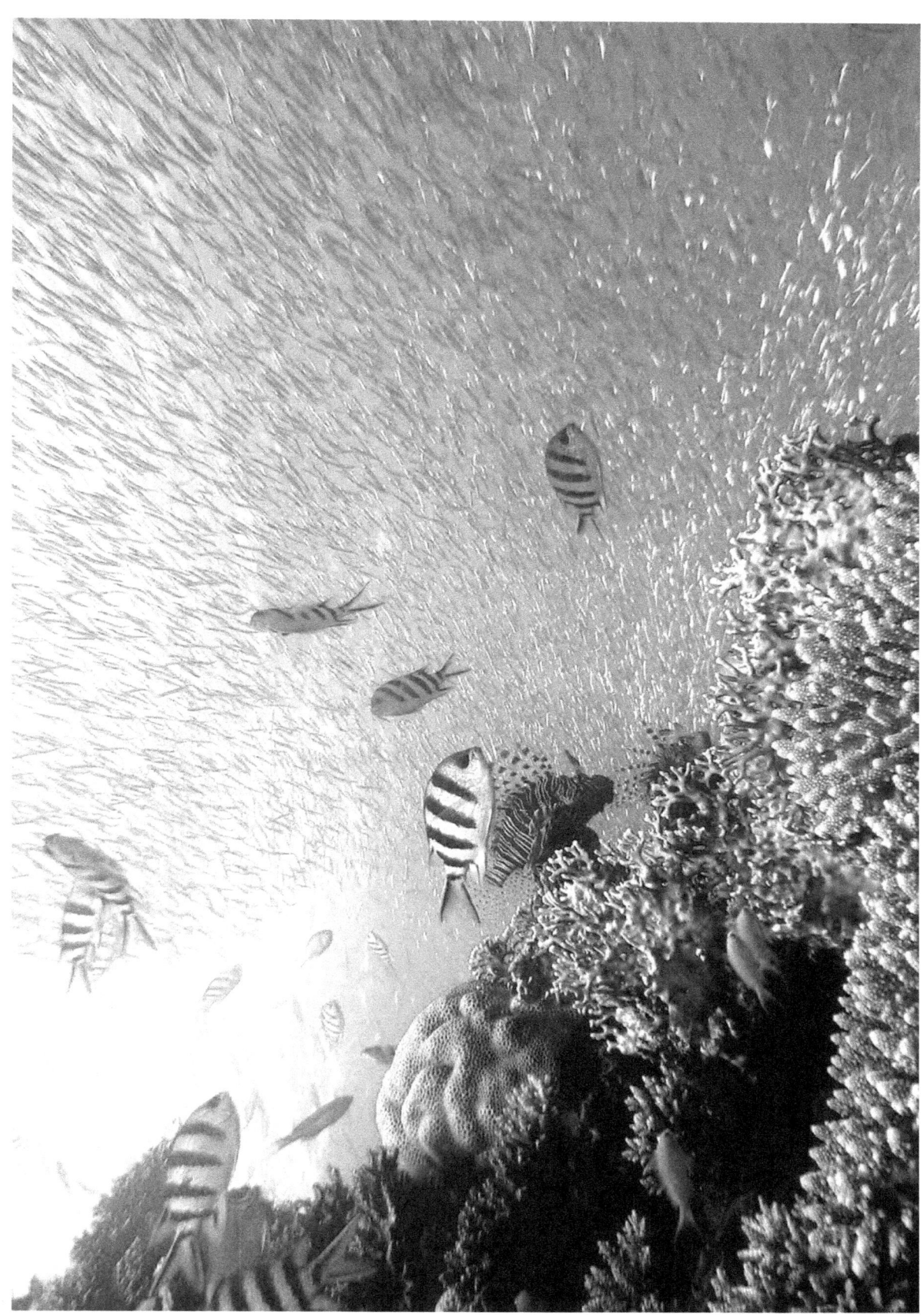

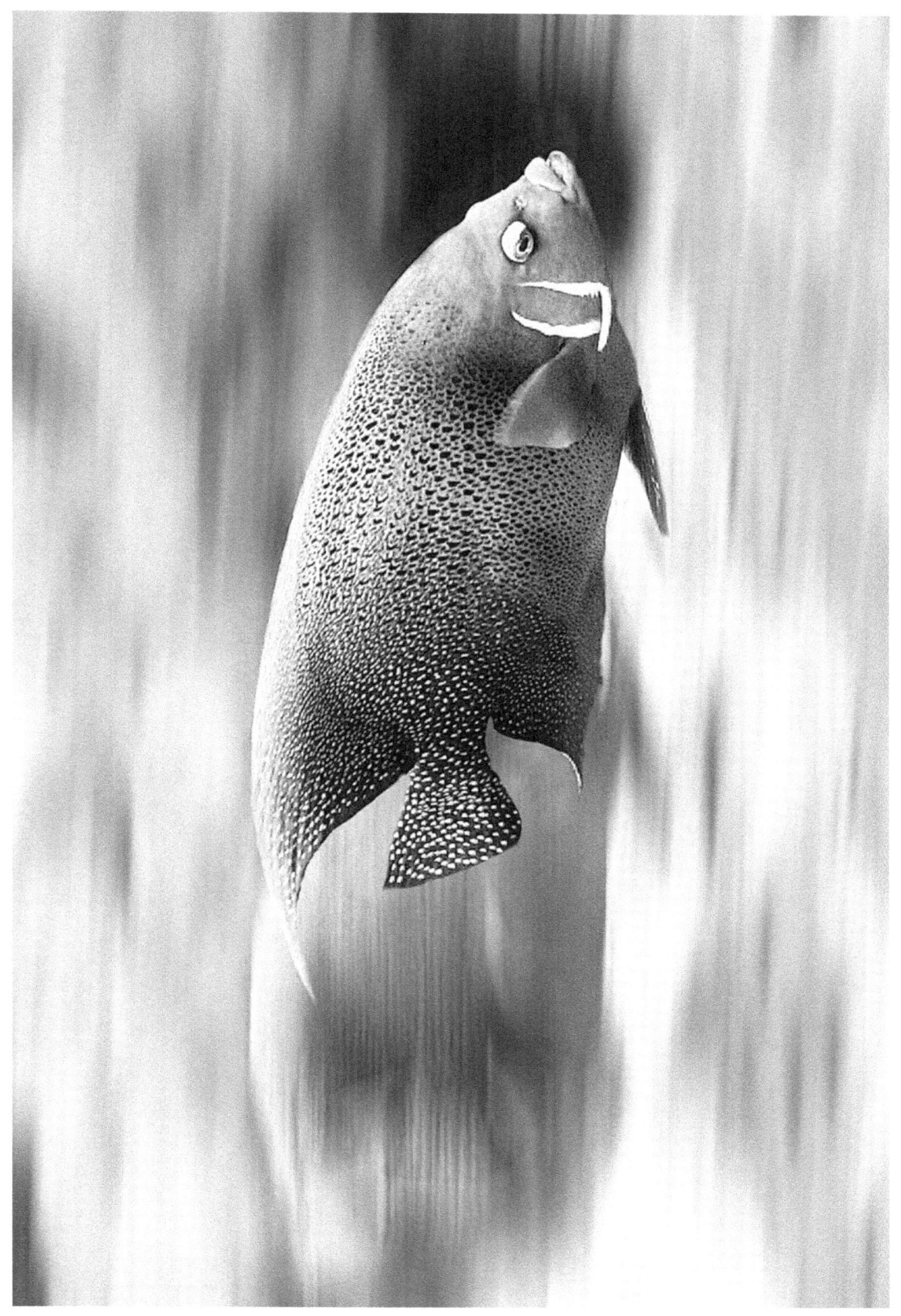

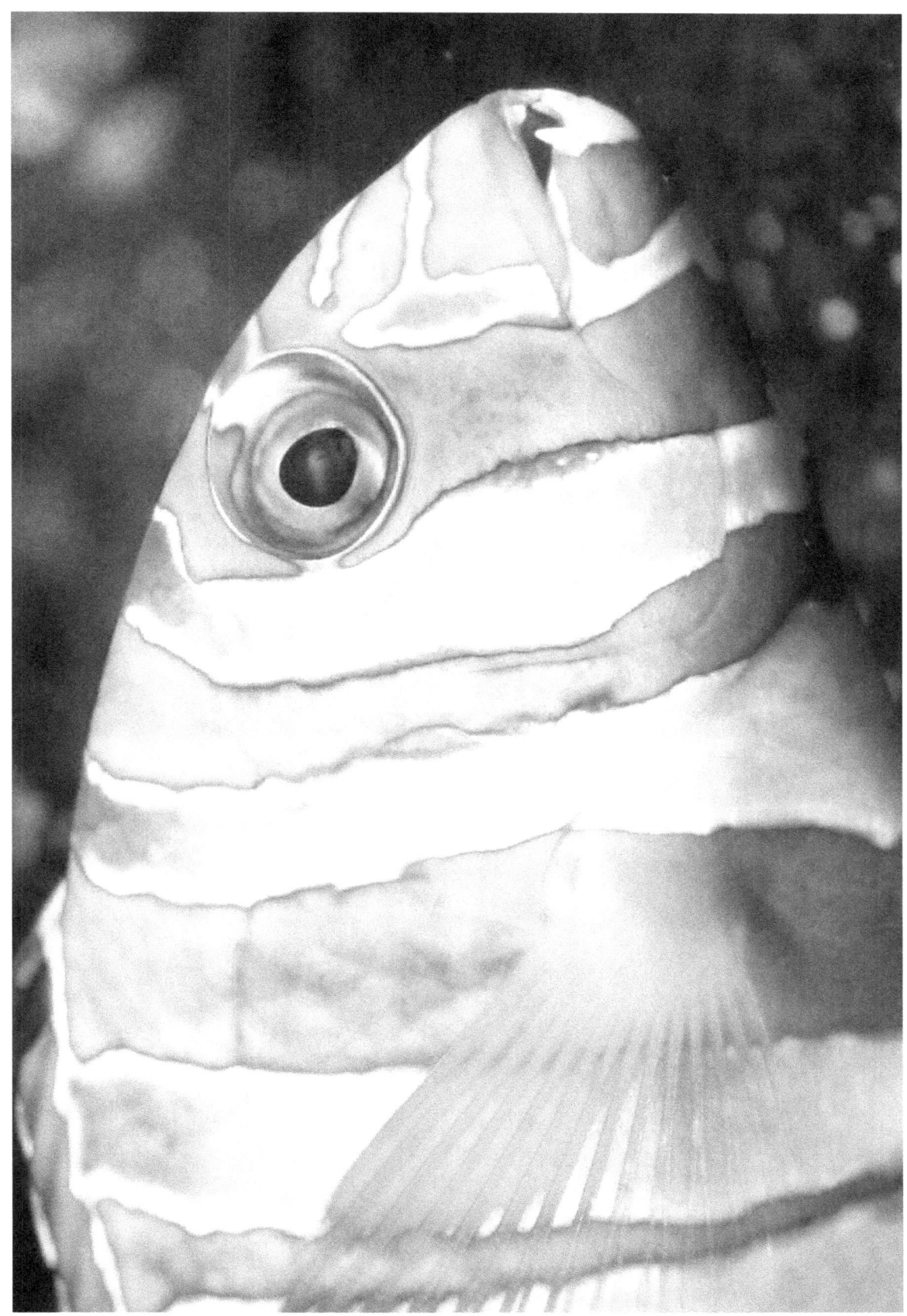

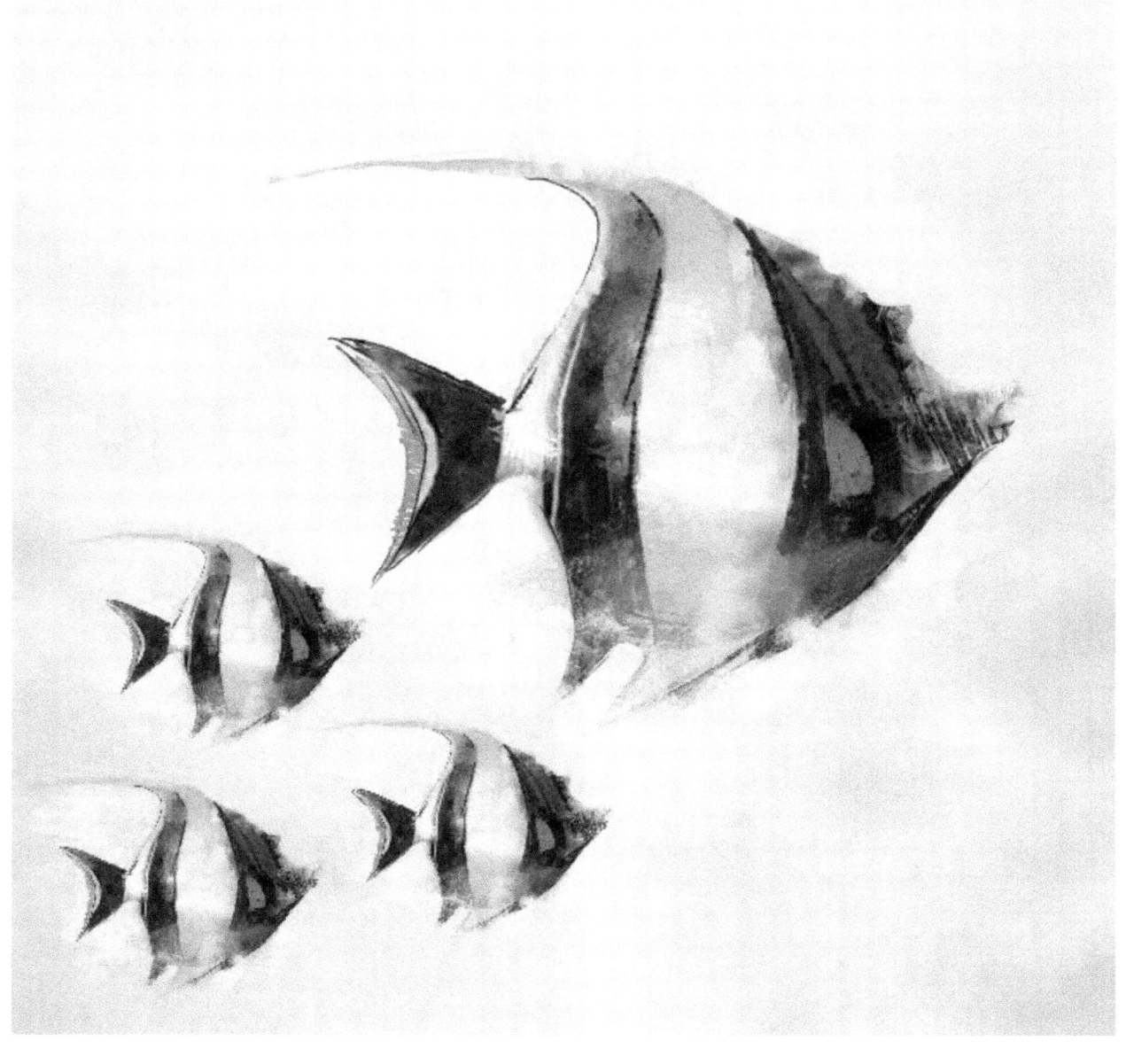

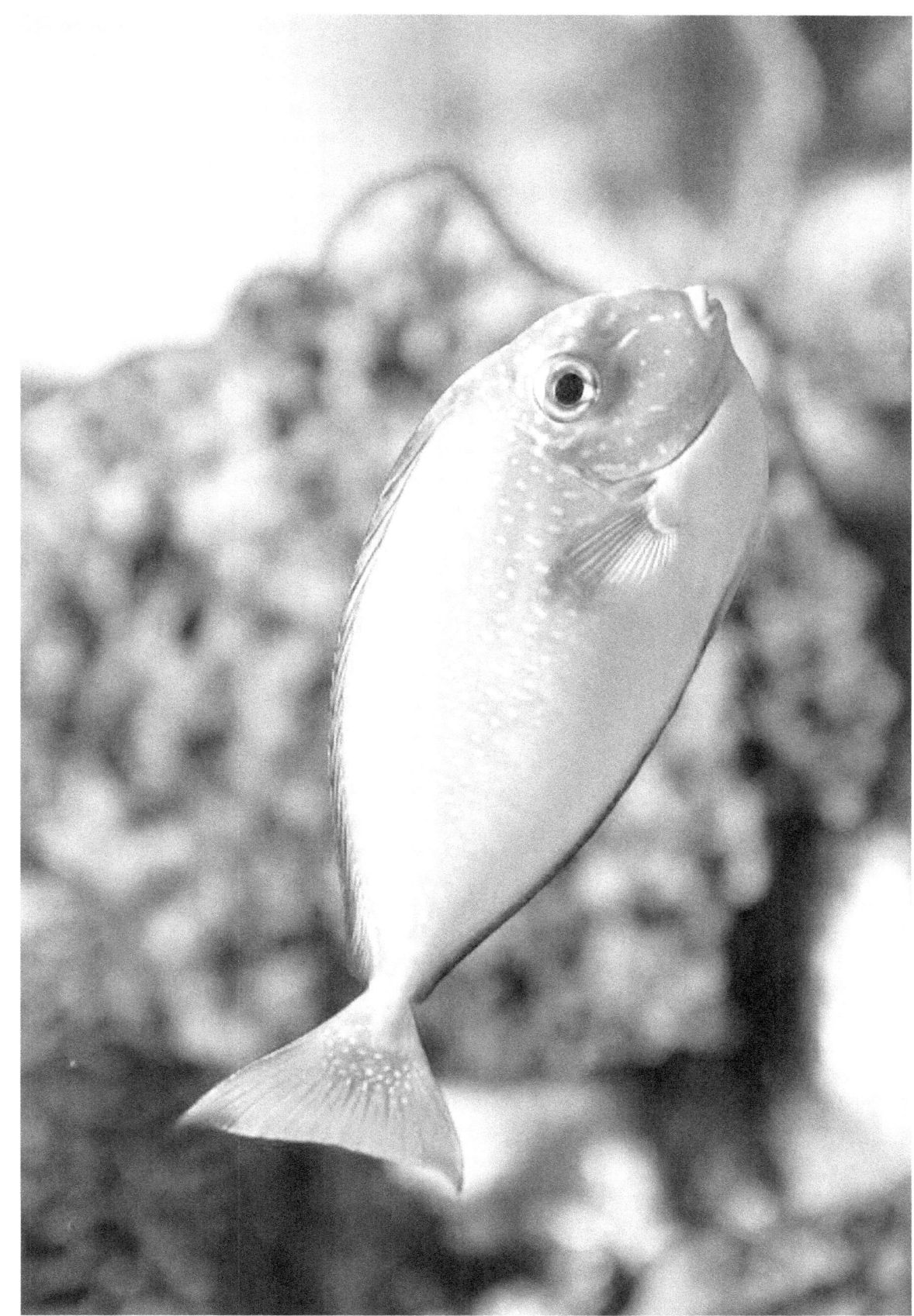

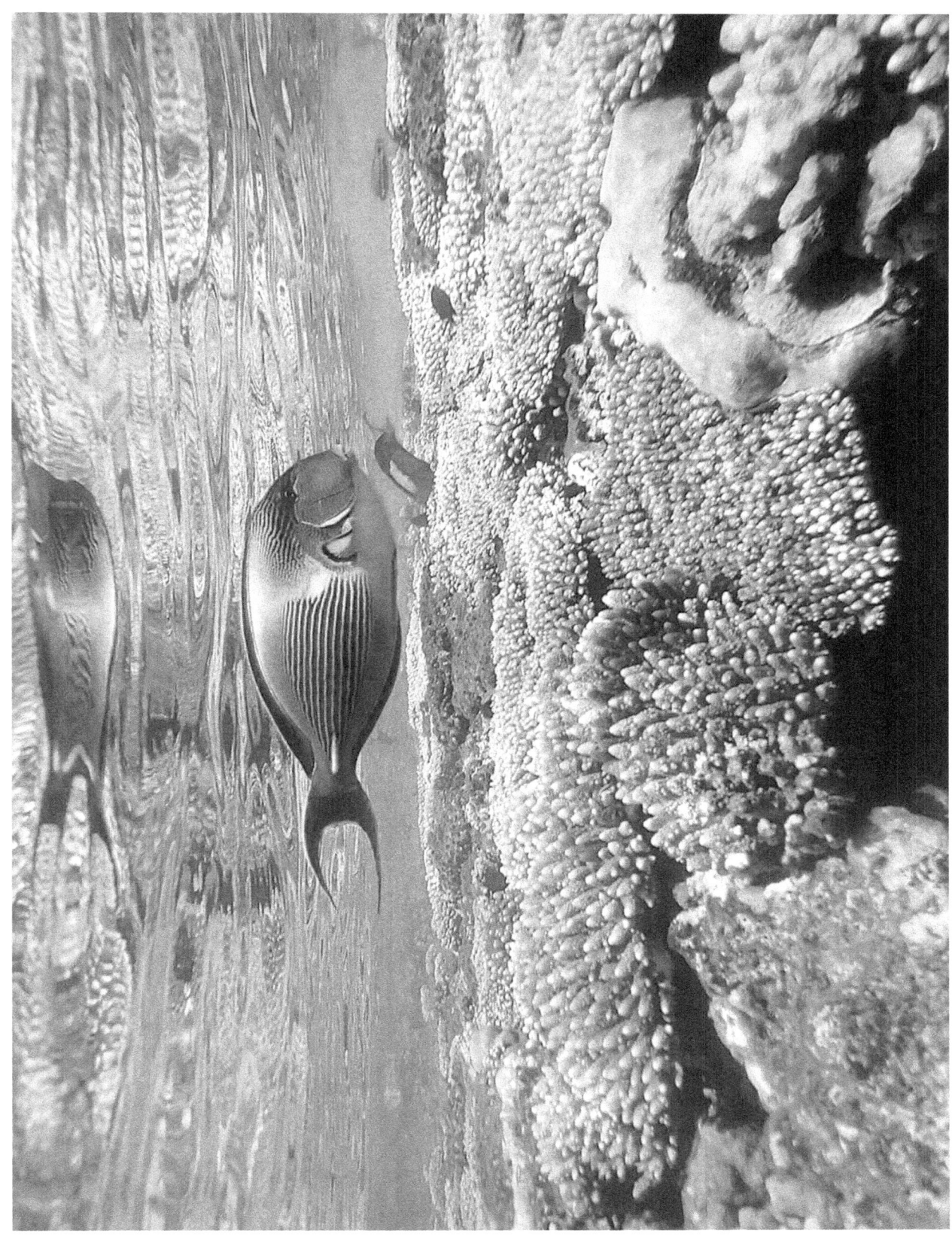

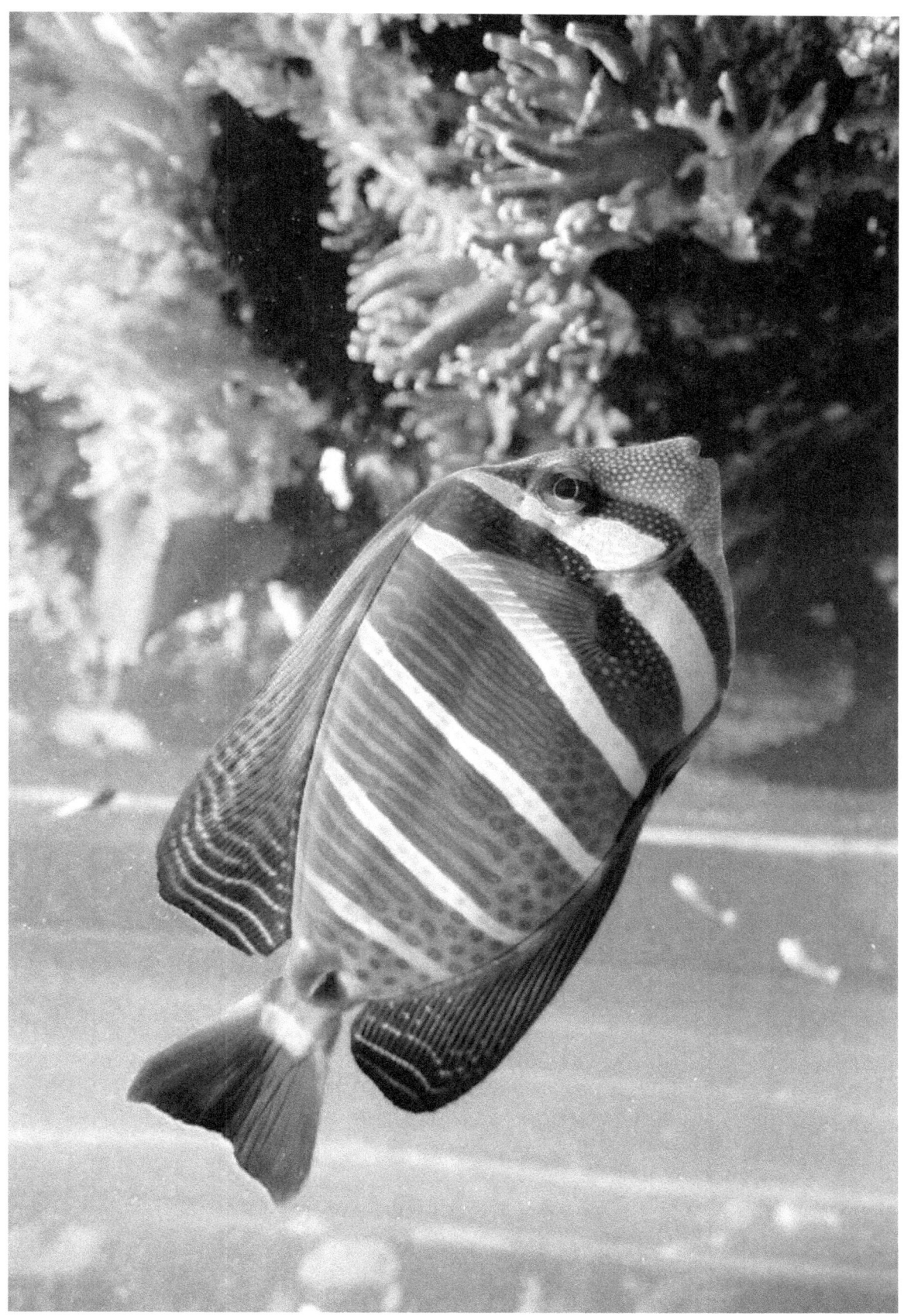

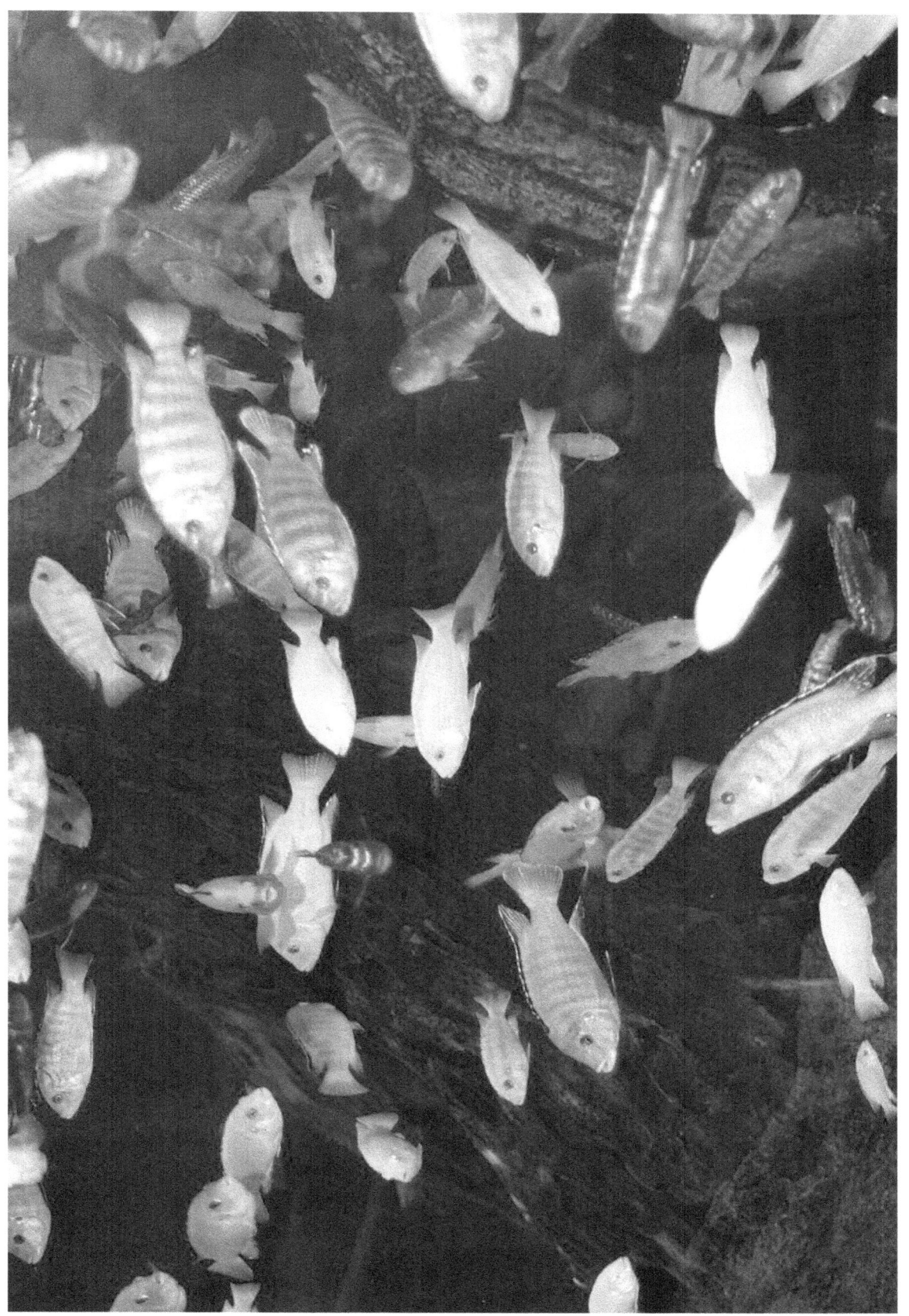

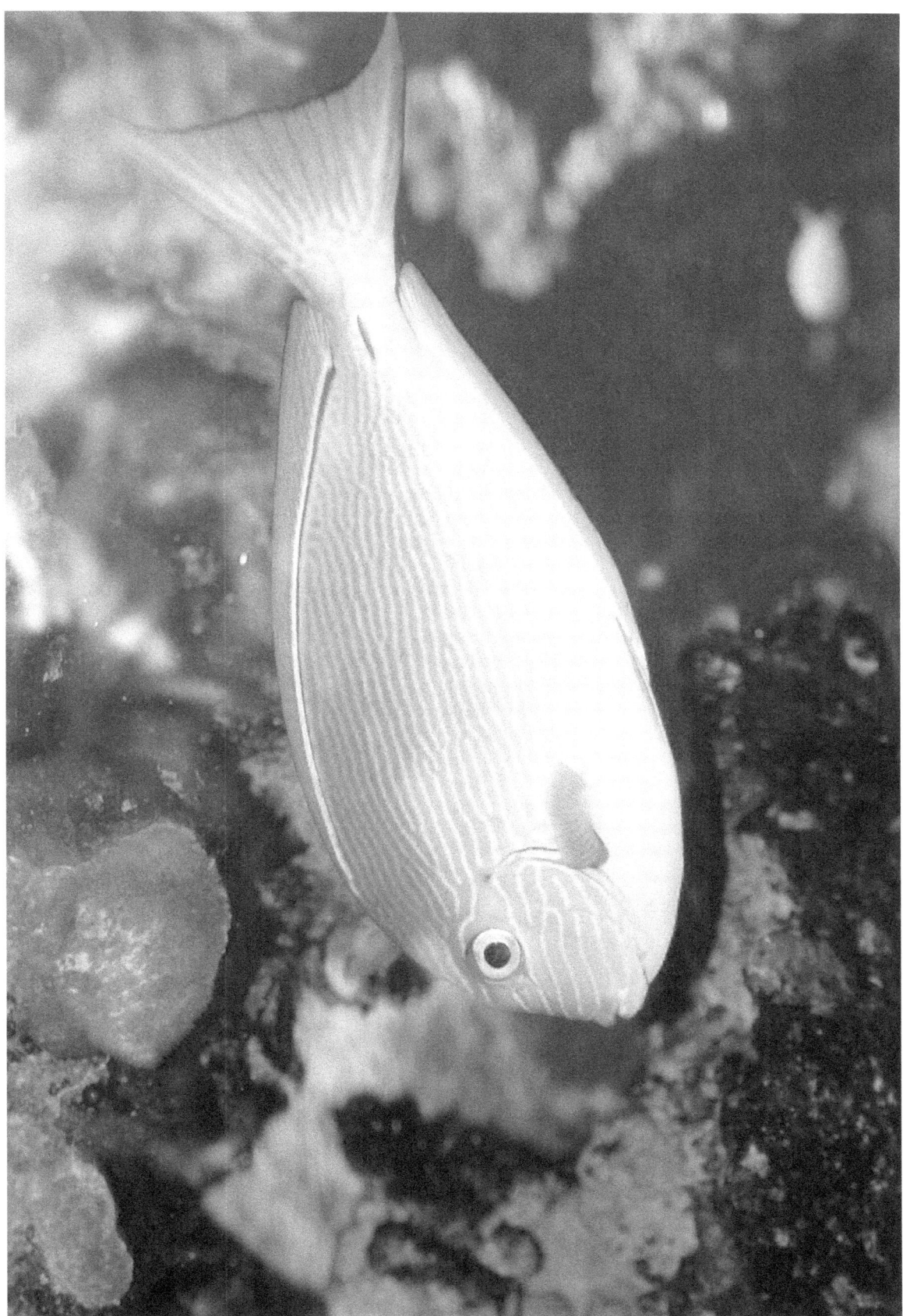

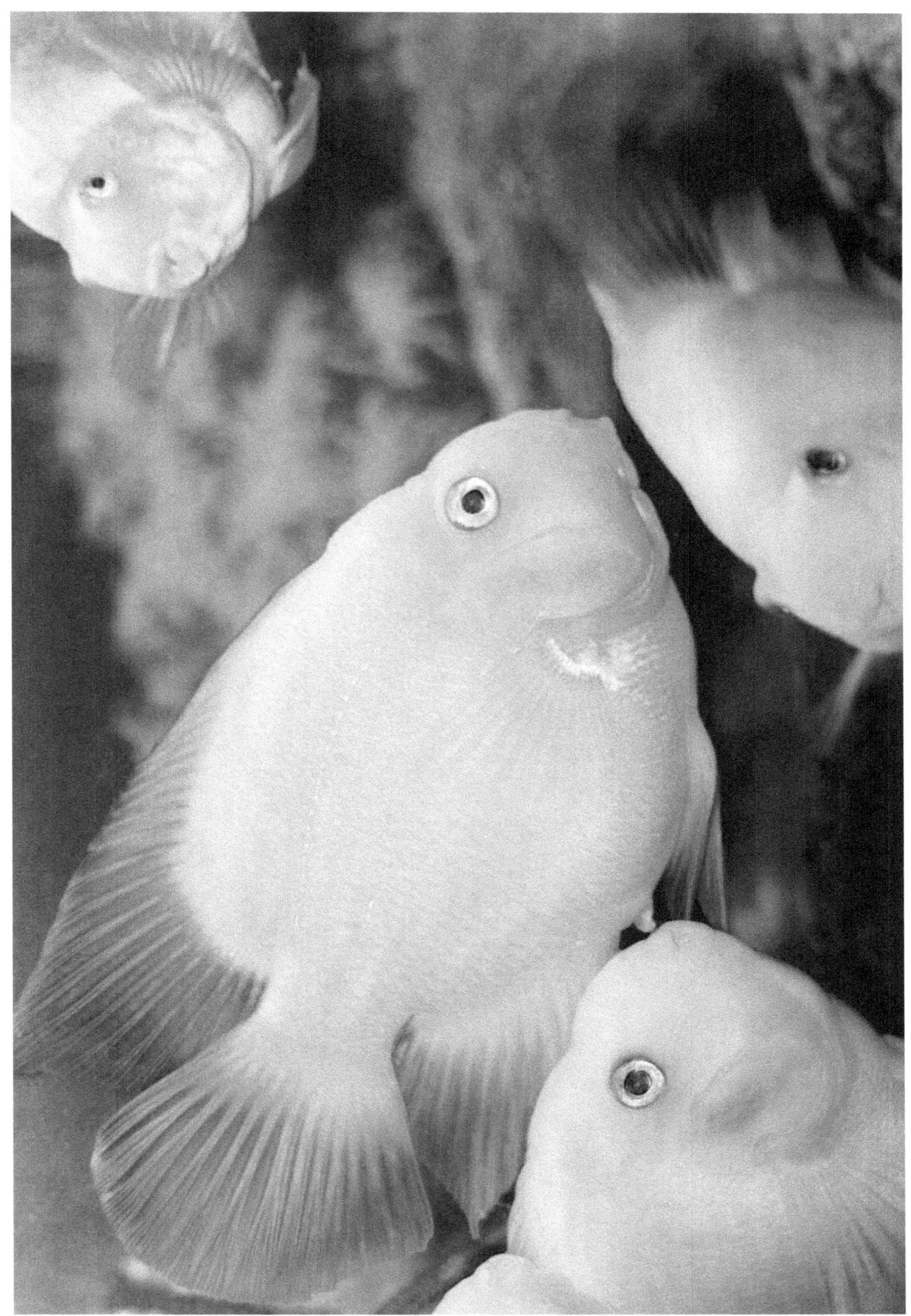

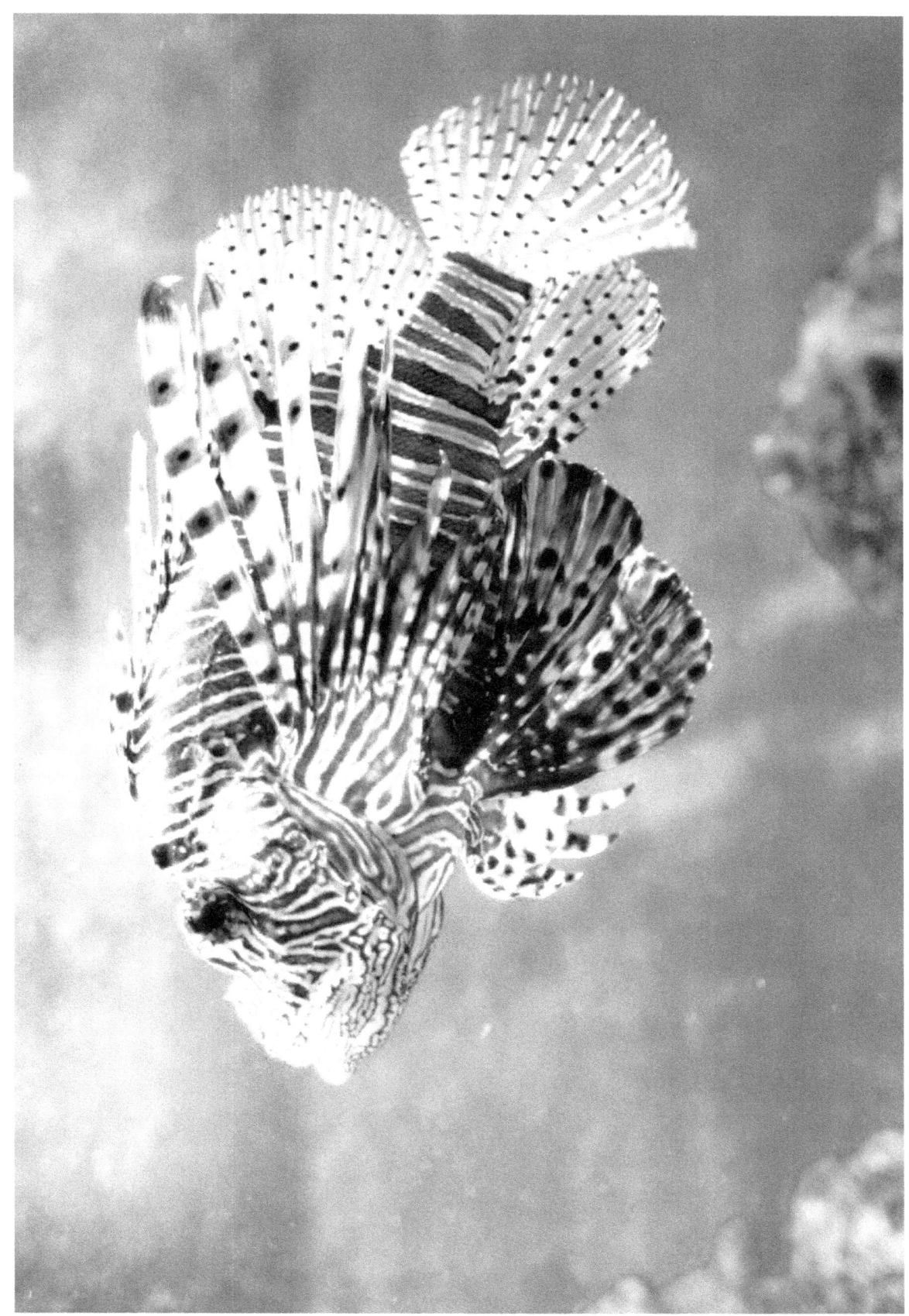

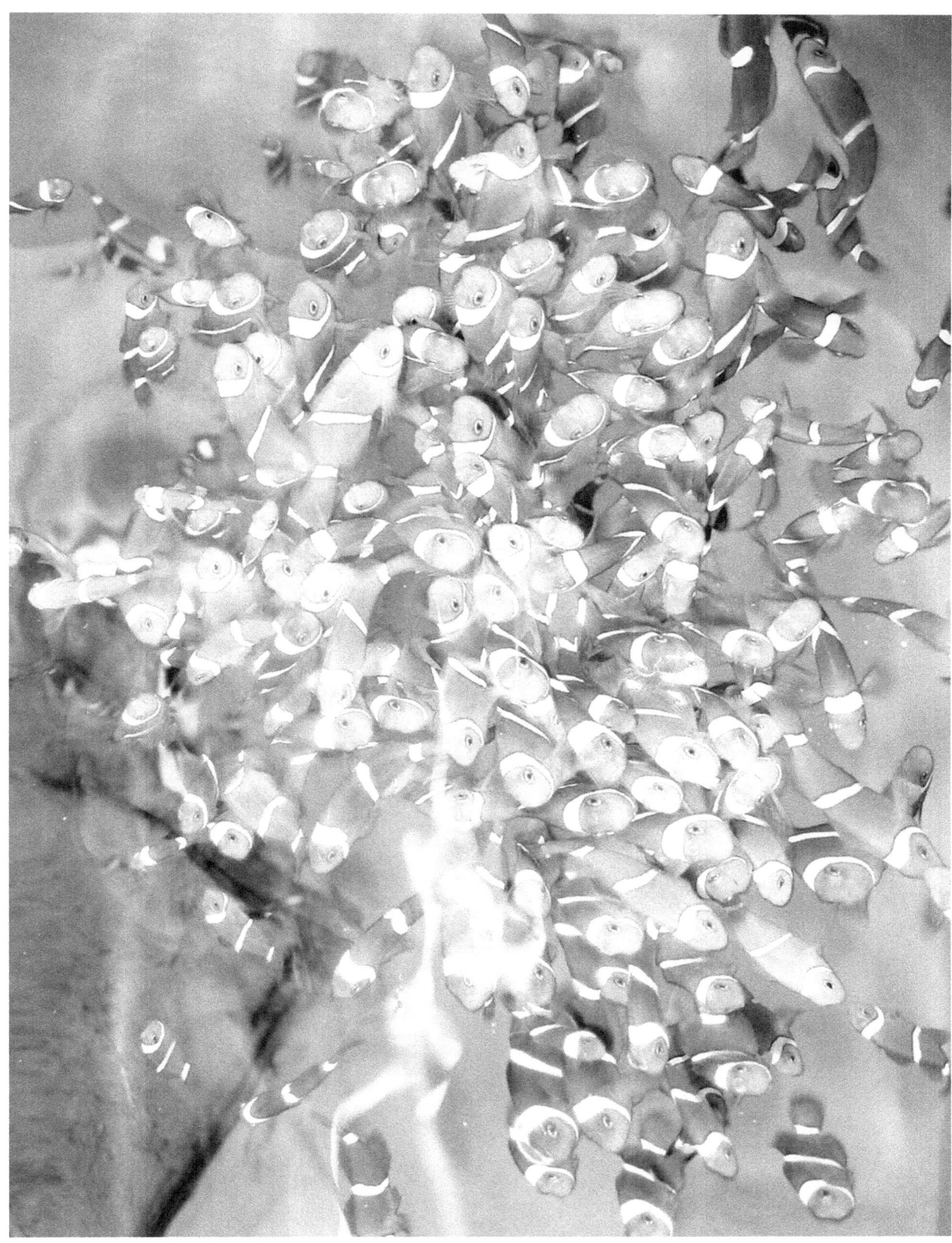

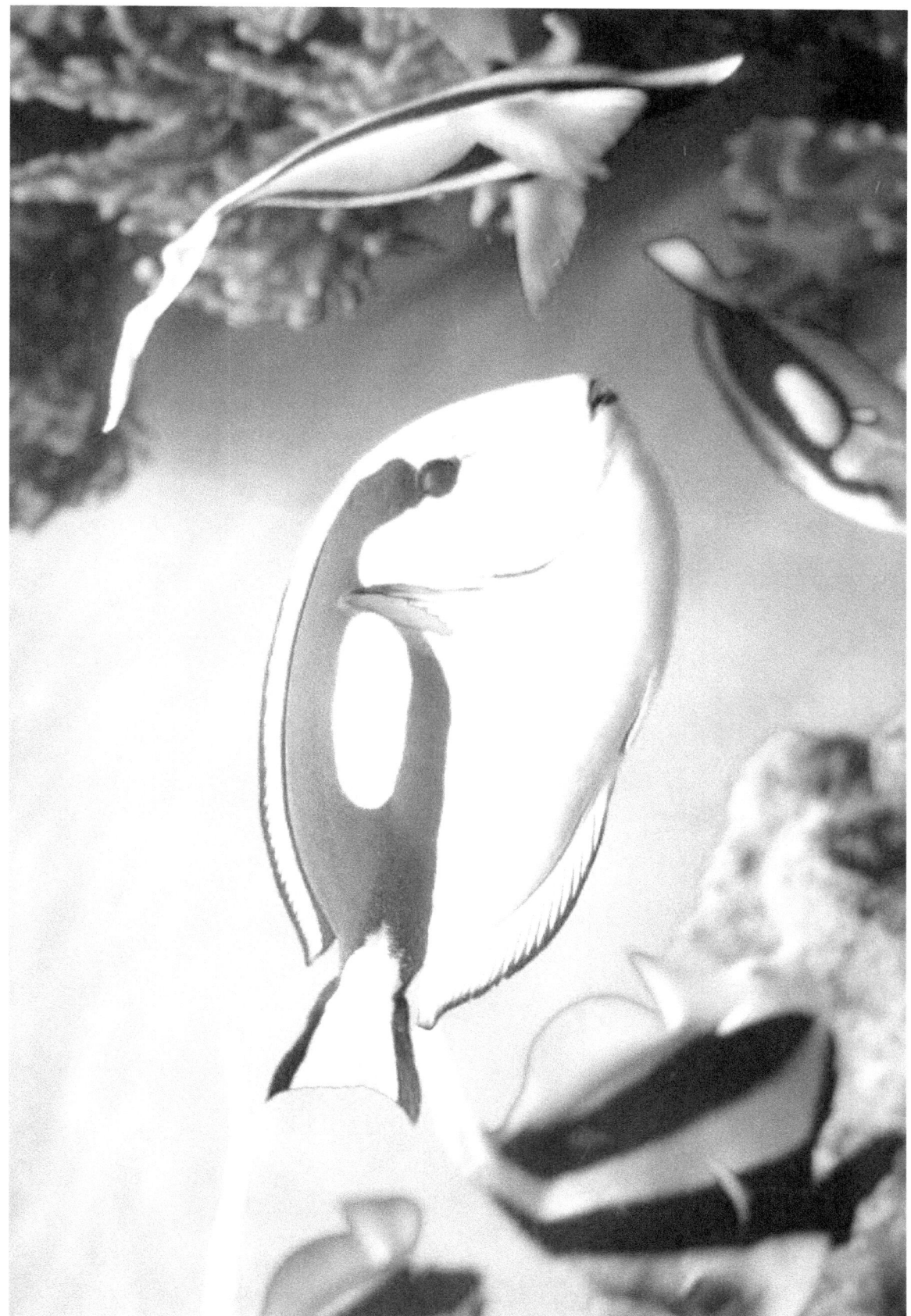

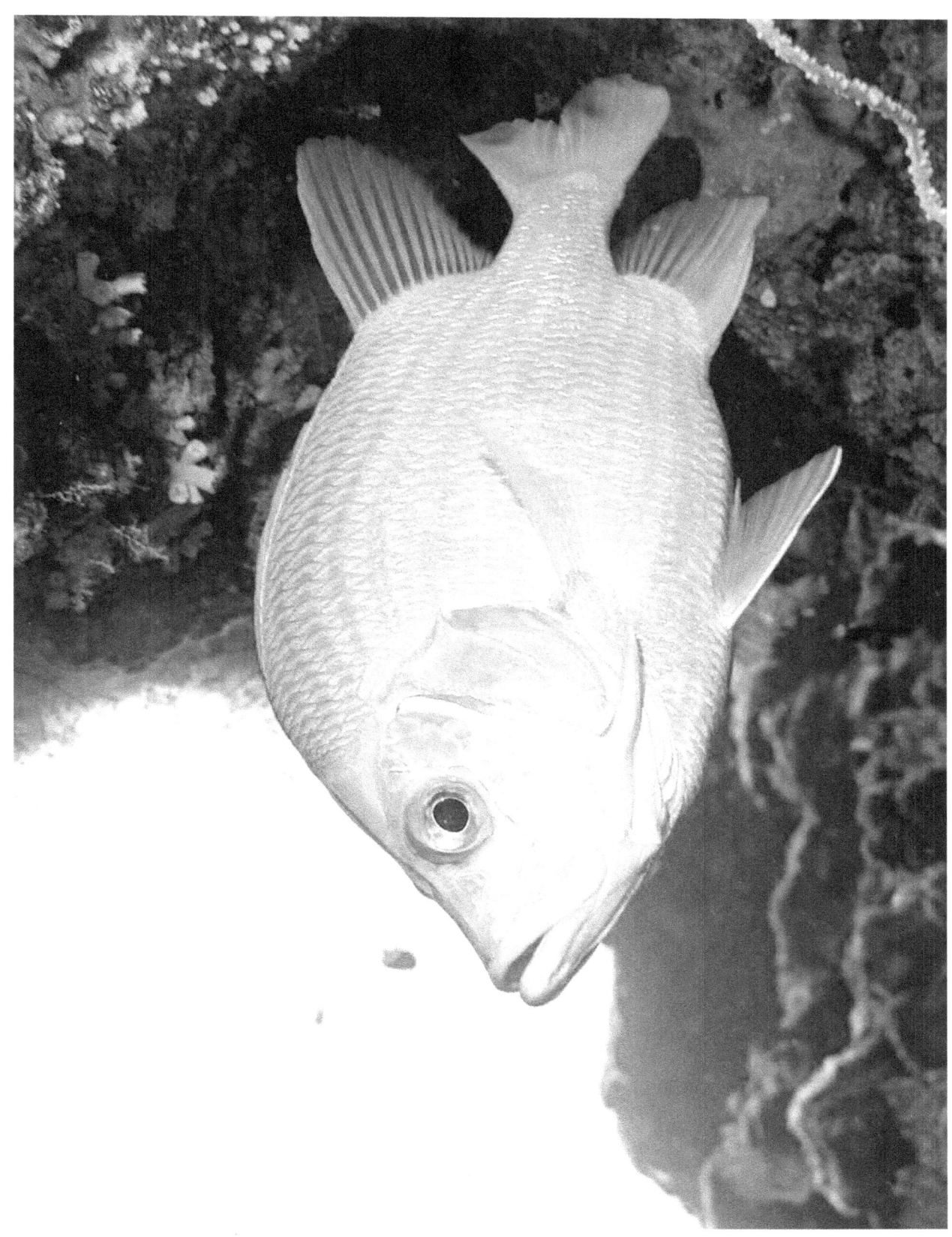

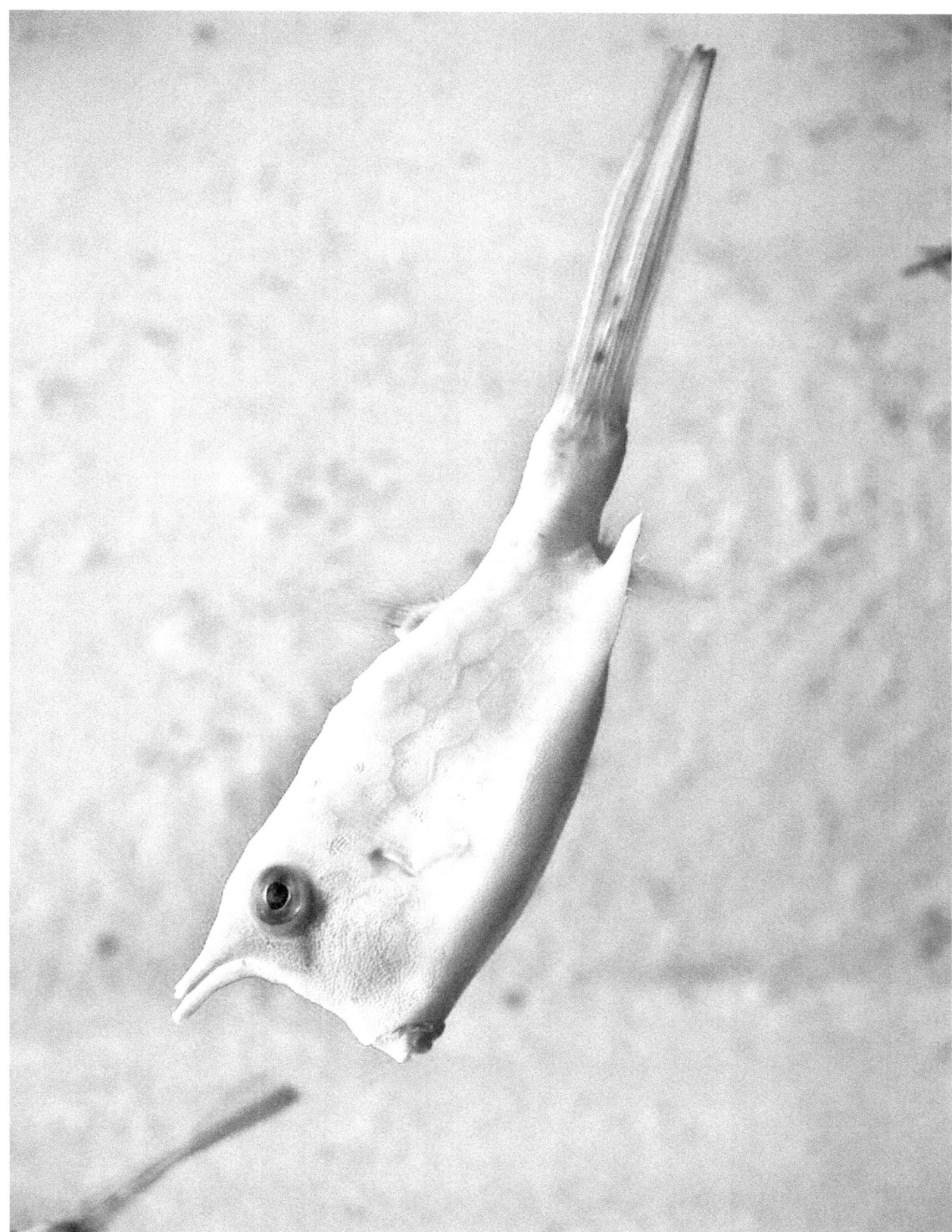

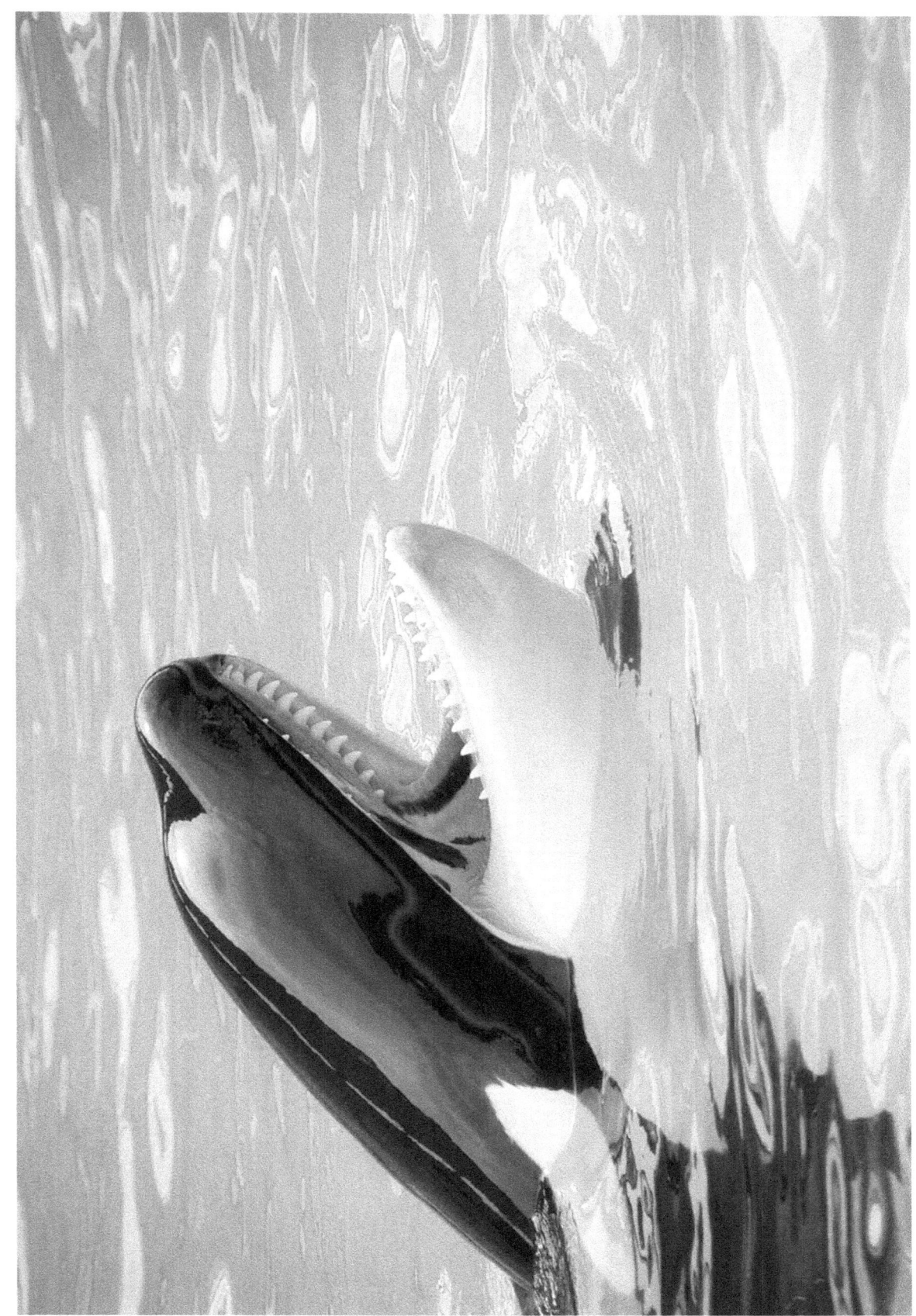

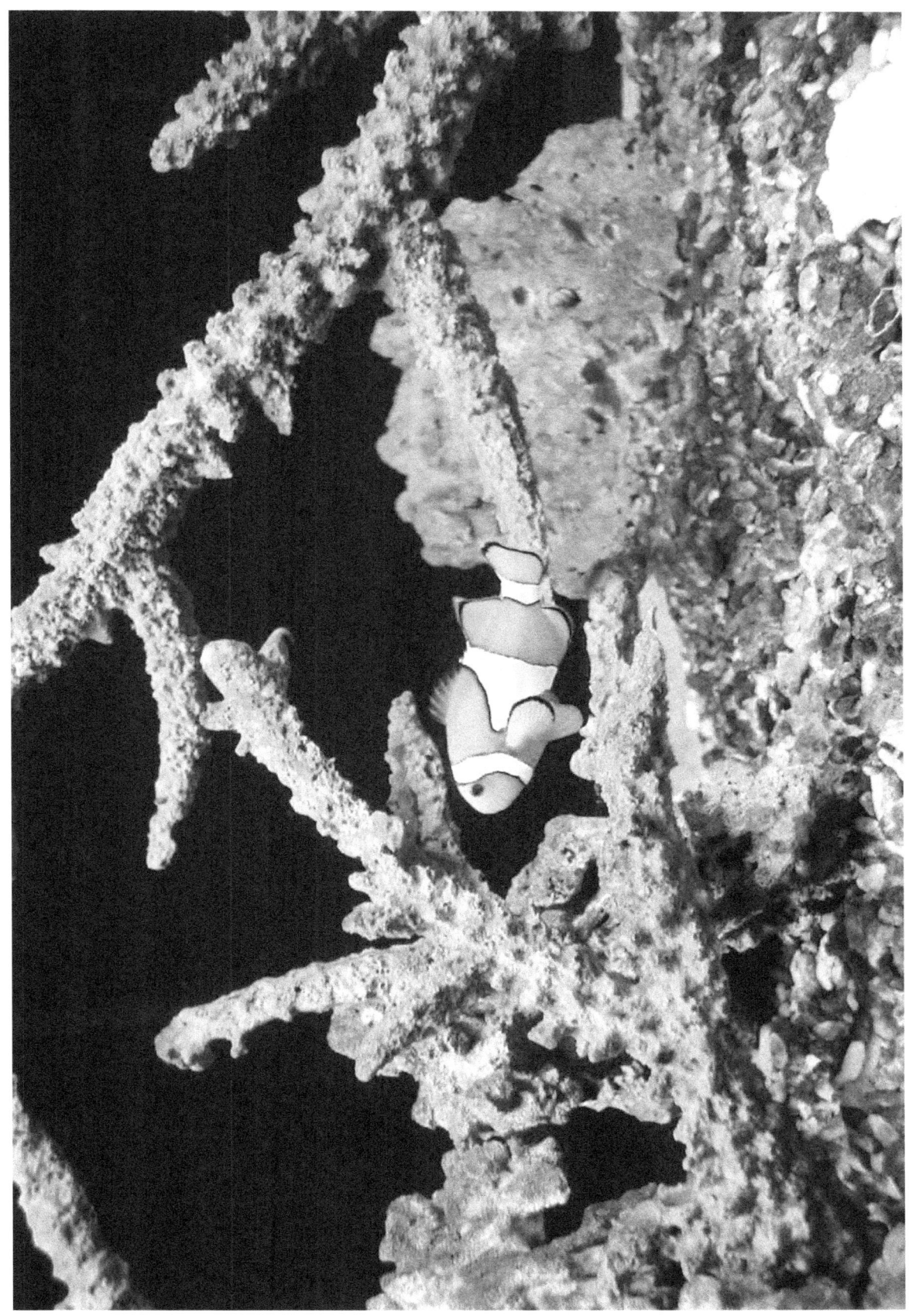

www.ingramcontent.com/pod-product-compliance
Lightning Source LLC
Chambersburg PA
CBHW080558190526
45169CB00007B/2818